Common Cognitive Biases
Examples and Challenges

SA Hale and Terry E. Hale

DEDICATION

To all the "unique and interesting" people
we've had to deal with... and the science behind
you.

CONTENTS

ACKNOWLEDGMENTS

The authors would like to thank all the scientists for their groundbreaking work in Cognitive Science, Neuroscience, and Behavioral Economics.

1 INTRODUCTION

Our brain has many ways of doing its job, including tricks and shortcuts that help it work efficiently. Studies have shown, however, that these tricks and shortcuts can lead to making predictable mistakes. Does your brain lie to you, or does it at least filter reality?

In this book, the authors will discuss some of the most common cognitive biases, how to identify these biases and heuristics cognition, how to potentially implement cognitive biases for use in your business and life arena and finally how to mitigate those cognitive biases. Research in Cognitive Psychology, Evolutionary Psychology, Cognitive Science, Neuroscience, Anthropology, and Human Reliability Engineering has identified over one hundred cognitive biases. These cognitive biases appear to be part of our evolutionary human brain and transcend cultures and generations. We can describe cognitive biases as a replicable pattern in perceptual distortion, inaccurate judgment, and illogical interpretation. Cognitive biases are the result of distortions in the human cognition that always lead to the same pattern of poor judgment, often triggered by a particular situation. Three main types of cognitive bias will be identified and discussed; Decision-making, Social Biases, and Memory Biases.

Because of the authors' particular area of interest, we believe that by developing an understanding of cognitive biases, cyber defenders will be able to be adaptable and resilient to meet the rapid and ever-evolving threat to our national security. We must understand our vulnerabilities because humans are the weakest link in sales, marketing, and security. Once understood, training, tools, and processes can be implemented in both the business world and the Information Technology world of Information Security, Information Assurances, and cyber warfare.

Keywords: cognitive biases, cognitive science, evolutionary psychology, neuroscience, behavioral economics, artificial intelligence, human reliability engineering, anthropology, paleoneurology, cognitive bias mitigation, decision-making biases, social biases, memory biases

2 DEFINITIONS OF COGNITIVE BIASES

Cognitive bias describes the inherent thinking errors that Homo sapiens make in processing information. Several have been verified empirically in the fields of psychology, neuroscience, and anthropology, while many others still need scientific investigation. Cognitive biases are thinking errors that prevent one from accurately understanding reality, even when confronted with all the needed data and evidence to form an accurate view of the situation. Many conflicts, errors, and mistakes are due to these cognitive biases that prevent people from coming to the same conclusions based on the same evidence. Cognitive bias is intrinsic to human thought, and therefore any systematic system of acquiring knowledge that attempts to describe reality must include mechanisms to control for bias, or it is inherently invalid.

Our mental system consists of interrelated items of assumptions, beliefs, ideas, and knowledge that an individual holds about anything concrete (person, group, object, etc.) or abstract (thoughts, theory, information, etc.). It comprises an individual's world view and determines how their abstracts, filters, and structures information received from the surrounding world.

Cognitive bias describes a repeatable pattern in perceptual distortion, inaccurate judgment, illogical interpretation, or what is broadly called irrationality. (Kahneman, Baron, Ariely) Cognitive biases are the result of distortions in the human cognition through patterns that lead to the same pattern of poor judgment, often triggered by a particular situation. Cognitive biases are influenced by evolution and natural selection pressure. Some are presumably adaptive and beneficial, for example, because they lead to more effective actions in given contexts or enable faster decisions when faster decisions are of greater value for reproductive success and survival. Others presumably result from a lack of appropriate mental mechanisms, i.e. a general fault in human brain structure, from the misapplication of a mechanism that is adaptive (beneficial) under different circumstances, or simply from noisy mental processes.

"Hidden ghosts" inside our heads are best adapted to help us with simple, concrete day-to-day problems, not highly mathematical or abstract problems such as solving differential equations or imaginary numbers. (Servitor)

Summary:

We would like to conclude our brief chapter on defining Cognitive Biases with relating links to more history and related information on Cognitive Biases.

https://en.wikipedia.org/wiki/Cognitive_bias

https://en.wikipedia.org/wiki/Cognitive_bias_mitigation

http://rationalwiki.org/wiki/Cognitive_bias

www.businessdictionary.com

3 THEORETICAL CAUSES OF COGNITIVE BIASES: GLOBAL VIEW

Despite several decades of research and experimentation, no comprehensive theory of what creates or causes cognitive biases has emerged. Grouping and categorizing cognitive biases are not straightforward so making listings of cognitive biases are a grab bag of heuristics and biases, which have no quantitative psychological, anthropological, human reliability engineering theory, research, or experimentation to support the grouping. (Dougherly) A 2012 Psychological Bulletin article suggests that at least eight seemingly unrelated biases can be produced by the same informational-theoretic generative mechanism that assumes noisy information processing during storage and retrieval of information in human memory. (Hilbert)

The following is taken from the work by Patrick Conroy of the University of British Columbia.

As Conroy states, "...since early 2011, I have been mining Cognitive Science, Evolutionary Psychology, Neuroscience, Behavioral Economics, Artificial Intelligence, Human Reliability Engineering, and related disciplines for a solution to the problem of biased reasoning." In addition, to searching for cognitive biases mitigation, Conroy has listed several currently debated possible causes of cognitive biases. What follows is a brief synopsis of his published findings.

System 1, System 2 Model

This is a simplification of the System 1, System 2 Model. This model, used by traditional psychologists and economists to characterize the human reasoning mechanism, suggests that you are your brain. This could be equated to a black-box computer complete with an operating system, data storage, and input and output devices inherited from past generations, and with application software being the rules of behavior we learn after birth.

Conroy and I agree that cognitive biases and their mitigation rely on the idea that our reasoning mechanism consists, approximately, of a combination of two components: 'System 1' and 'System 2.' The definition for System 1 is that it consists of heuristic shortcuts developed over evolutionary time and implemented in certain brain cortices as a kind of Heuristic Toolkit. These heuristics are invoked automatically by stimuli from both the outside world and our internal neural system; they are massively parallel and near-instantaneous in their operation.

Many people believe as Conroy does, that cognitive biases in System 1 cannot be 're-programmed,' are invisible to our conscious mind, and their main job is to provide 'advisories' to System 2. Evidence that supports this theory is found in evolutionarily psychology, anthropology, and paleoneuology. Conroy continues that, by contrast, System 2 is the evolutionarily modern component of our reasoning mechanism, which operates serially, slowly, and (partly) visible to our conscious minds. This component 'simulates' the future consequences of System 2's advisories, rejects clear 'losers,' redirects the projected stimuli for the rest back to System 1, and at the end of several such iterations, selects one advisory as the judgment, decision or choice to run with. This view is one involving a fair amount of hand-waving but is sufficiently backed by evidence from neuroscience and other disciplines that it can support the kind of research program defined above.

Emotional Agency vs. Rational Agency

Neo-classical Economics views people as <u>rational agents</u> making choices that maximize a personal Utility Function without empathy for effect on others and using formal analytical methods (for example, mathematics, logic, probability, statistics, system dynamics, etc.) to determine the values used in the Utility Function. This view effectively casts us all as machine-like psychopaths, motivated by selfishness and constrained only by external punishments. By contrast, the authors hold the view that we make our choices on the basis of how well they serve our well-being and that of the social groups we belong to, that our reasoning is 'fast and frugal' based largely on pattern recognition rather than exhaustive, high-bandwidth computation, that the 'valuations' we put on various aspects of well-being reflect our human needs and motivations (for example, Maslow's Hierarchy of Needs) and that the 'weights' we attach to these valuations are primarily emotional in nature. This view casts us all as social beings, motivated and constrained by an array of intrinsic social norms. (Convoy)

Heuristics vs. Flaws

Traditionalists view the 'reasoning errors' reliably produced in highly controlled laboratory experiments as evidence of flaws in the machinery of the brain. Reflecting their view of human reasoning as a computer running a vast array of formal analytical tools and methods based on fixed programming, the only way to prevent such errors is to re-program the brain with the correct software, presumably through intensive training in formal analytical methods. Essentially, the traditionalists' mitigation protocol would be universal and simple, regardless of the situation, i.e. become a rational agent in a sense assumed by neoclassical Economics. In the authors' view, this position is beyond simple; it is dangerously one-dimensional and tragically simple-minded.

Conroy subscribes to the view that the 'reasoning errors' reported in endless studies of Cognitive Bias arise from the inappropriate application of the Heuristic Toolkit in the System 1 component of our reasoning mechanism. Such inappropriateness results from mismatches between the reasoning challenges and social environments that originally produced our Heuristic Toolkit over evolutionary time, and the very different challenges and environments we face today. Cognitive Bias Mitigation in this view is a process for invoking an appropriate mix of System 1's Heuristic Toolkit and System 2's Prospection to match better the reasoning challenges at hand and the real-world social environment in which they meet.

Summary:

The above were the "Big Picture" or Global models for cognitive biases. The next chapter will describe more specific views of the causes of cognitive biases.

4 THEORETICAL CAUSES OF COGNITIVE BIASES: SPECIFIC CAUSALITIES VIEW

Bounded rationality is the idea that in decision-making, the rationality of individuals is limited by the information they have, the cognitive limitations of their minds, and the finite amount of time they have to make a decision. (Elster) ...one seeking a satisfactory solution rather than the optimal one. (Gigerenzer)

Attribute substitution is a psychological process thought to underlie some cognitive biases and perceptual illusions. It occurs when an individual has to make a judgment (of a *target attribute*) that is computationally complex and instead substitutes a more easily calculated heuristic *attribute*. (Newell) This substitution is thought of as taking place in the automatic, *intuitive* judgment system, rather than the more self-aware *reflective* system.

Hence, when someone tries to answer a difficult question, they may answer a related but different question, without realizing that a substitution has taken place. Individuals can be unaware of their

11

biases, and why biases persist even when the subject is made aware of them.

Cognitive dissonance is a discomfort caused by holding conflicting cognitions which are ideas, beliefs, values, emotional reactions simultaneously. In a state of dissonance, people may feel surprised, dread, guilt, anger, or embarrassment. (Festinger) The theory of cognitive dissonance in social psychology proposes that people have a motivational drive to reduce dissonance by altering existing cognitions, adding new ones to create a consistent belief system, or by reducing the importance of any one of the dissonant elements. (Festinger)

The availability heuristic is a mental shortcut that uses the ease with which examples come to mind to make judgments about the probability of events. The availability heuristic operates on the notion that "if you can think of it, it must be important." The availability of consequences associated with an action is positively related to perceptions of the magnitude of the consequences of that action. (Tversky) For example, if someone asked you whether your college had more students from Colorado or more from California, under the availability heuristic, you would probably answer the question based on the relative availability of examples of Colorado student and California students. If you recall more students that come from California that you know, the more likely you will conclude that more students are from California. (Matlin)

Summary:

In closing this section, there are numerous reasons, theories, and debates as to why Homo sapiens have a heuristic toolkit and cognitive biases. No one model has won the lead. The authors personally like the System 1, System 2 viewpoint. The question remains as to why cognitive biases are cross-cultural, cross-racial, cross-social-economics status.

5 OVERVIEW OF THE THREE DISTINCT CATEGORIES OF COGNITIVE BIAS

Decision-Making Biases

"These influence us when we are about to make an important decision, whether it is to buy something, claim something in a discussion, to choose where or how to stand in a debate, etc." (Servitor) Continuing our definition of decision-making from Wikipedia, many of these biases affect belief formation, business and economic decisions, and human behavior in general. They arise as a replicated result to a particular condition when confronted with a given situation.

Finally, "Cognitive biases steer our decision making in subtle-but-definite ways. They might make us a little more Pollyanna-ish, a bit more self-centered, a bit more negative, somewhat more what-have-you-done-for-me-lately, or a little more something else. Little, but measurable, pushes to our decision in reactive directions." (Benson)

Social (Attribution) Biases

"These influence us when it comes to understanding ourselves, our actions in relationship to other people and their actions." (Servitor) An attribution bias affects the way we determine who or what was responsible for an event or action (attribution). It is a cognitive set that may interfere with social interaction. Wikipedia continues by stating that "Attribution biases," social biases, "typically take the form of *actor/observer differences*: people involved in an action (*actors*) view things differently from people not participating (*observers*).

These discrepancies are often caused by asymmetries in availability, known as salient. Taylor and Fiske (1975) concluded that "the more conspicuous a potential causal agent is, the greater the causal role observers will assign to the agent in accounting for particular outcomes of events." (Jones)

In the book Tradecraft Primer by the CIA, Attribution biases (Social) are observed as "Behavior of others is attributed to some fixed nature of the person or county, while our behavior is attributed to the situation in which we find ourselves" A "bias in perceiving causality."

Memory biases

This category of biases influences what, how, when, where, and why we remember certain memories the way we do. (Servitor)

Wikipedia defines memory bias as a cognitive bias that either enhances or impairs the recall of memory. Either the chances that the memory will be remembered or the amount of time it takes for the memory to be remembered. Of the content of a recalled memory can also be affected.

As with all Cognitive Biases, there are many types of memory bias; "...cues such as the ease with which a scenario or event comes to mind, and the frequency of such events in memory may significantly influence our estimation of their actual likelihood". (Jones) Aamodt and Wang established that your memory of the past is unreliable and your perception of the present is highly selective.

Summary:

Most authors divide cognitive biases into sub-categories to better understand and map to human brain or personality under study. The above categories were one attempt at grouping cognitive biases this is by no mean the definitive grouping. Others, as you will see later in the book have grouped cognitive biases into four categories; Social biases, Memory biases, Decision-making biases, and Probability or Belief biases. Ever how you sub-divide cognitive biases is interesting, but the real fun is reading, observing, and playing with real examples.

6 COMMON COGNITIVE BIASES WITH EXAMPLES THE AUTHORS HAVE OBSERVED AFFECTING PEOPLE IN THEIR EVERYDAY LIFE

Ambiguity Effect: The authors believe that this bias is one of the cornerstones of Behavioral Economics. From Wikipedia; The ambiguity effect is a cognitive bias where decision making is affected by a lack of information, or "ambiguity." The effect implies that people tend to select options for which the probability of a favorable outcome is known, over an option for which the likelihood of a favorable outcome is unknown.

Category: Belief and Decision Bias

Example:
- Salespeople often show a very expensive item first, making all other options seem more affordable.

- Most people would choose a regular paycheck over the unknown payoff of a business venture.
- "If a board of directors is deciding whether to keep with the same strategy that is continuing to lose steam or take a chance on a new one—they're likely to feel an urge to stick to what they've seen and understand." (Anderson)
- "The devil you know." Think about ordering a familiar dish at a new restaurant versus ordering something that sounds great but is new to your palate.

Related to: Loss Aversion

Anchoring Bias: A tendency to rely on or make a decision based on a past reference or a single trait, or a single piece of information, or the first option presented to you. "Whether for good or for bad, the first piece of information we receive about a person or situation will color our overall perception." (Brudner) All further information about an individual or situation is judged by the initial meeting or information details; this is the anchor.

Category: Belief and Decision Bias

Example:
- All the stuff about how first impressions matter is right. How you introduce yourself and your product, your security product, policy, or procedures does in fact matter.
- The first snippet of information that you present to someone should set a positive tone, a win/win for the listener. Say what you want up front. Don't build up to it, as we do in a scientific manner. Present what is your desirable outcome first then add supporting material as need.
- Giving the specific information at the top of the discussion ensures that you and management are thinking toward the same goal from the start.

Also known as Focalism

Availability effect: is a mental shortcut that relies on immediate examples that come to a given person's mind when evaluating a particular topic, concept, method or decision.

Category: Belief and Decision Bias

Example:

- You have been watching the latest news reports about the number of car thefts around the country. Because you lack the time or resources, you make a judgment that this is happening in your community. You make a decision that car thefts are more common than they are in your area. The availability heuristic allows people to make an immediate decision, Many time these conclusions are incorrect.
- Like other biases, the availability bias can be useful at times. But it tends to lead to problems and errors in judgment. Examples of reports such as child abductions, airplane accidents, and train derailments have a tendency to lead people to the belief that such events are the norm.
- After hearing reports about people losing their jobs, you start ruminating that you're in trouble of losing your position. You start lying awake in bed each night worrying that you are about to be fired.

- After watching "Shark Week," you begin to think that incidences of attacks are a common occurrence. Now, you have a trip to the beach, and you refuse to go swimming in the ocean because of the high number of attack.

Related to or also known as illusory truth effect, or truth effect or the illusion-of-truth effect

Bandwagon Effect: The tendency to do (or believe) things because many other people do (or believe) the same.

Category: Belief and Decision Bias

Example:
- Fads of all types.
- Voting for the person who appears to be winning even if you don't like them but everybody you know are voting for the person or the option polls tell you that a particular candidate is leading.
- Buying a product just because everyone is buying that product.
- You can observe this bias during a bull market

Related to: Groupthink, Herd Behavior

Confirmation Bias: The tendency to search for or interpret information in a way that confirms one's preconceptions. In conversations the impulse to lock onto a part of a talk that confirms your existing beliefs, values, and perceptions.

Category: Belief and Decision Bias

Example:
- Scientific ideas (vaccinations, climate changes, etc.)
- political parties
- sport fans
- Imagine that you're researching a potential product. You think that the market is growing, you find information that supports this belief. Because of your research. You decide that the product will sell quite well, with backers and an extensive marketing campaign you launch your product...to a total failure. Everything is wrong; market did not grow, fewer customers are generated, the product tanks. It is an example of interpreting the product growth in a way that reinforces your preconceptions.

Curse of Knowledge: When better-informed people find it difficult to think about problems from the perspective of lesser-informed people. "The curse of knowledge is a cognitive bias that occurs when, in predicting others' forecasts or behaviors, individuals are unable to ignore the knowledge they have that others do not have, or when they fail to disregard information already processed. Things become second nature, so we forget about details, leaving others in the dark.

Category: Belief and Decision Bias

Example:
- People in technical fields or specialties filed
- Where to begin when teaching a language?
  - An example of the curse of knowledge is demonstrated in a classroom setting, where teachers, or subject experts, have difficulty teaching novices because they cannot put themselves in the position of the student. A brilliant professor may no longer remember the difficulties that a young student may be encountering when learning a new subject.
- Telling someone how to build a watch when they just ask for the time.

Decoy Effect: The Decoy Effect comes into play when consumers are choosing between two choices and have a preference, but change their preference when a third strategically priced option is offered.

Category: Belief and Decision Bias

Example:
- When you have two options, users will typically pick the cheaper options; this is not the option you as the salesperson wanted. How do you encourage the buying of the more expenses option and have the buyer believe it is their idea? This is a perfect situation for the decoy effect! Now, you add a third option the decoy; it is priced close to the more expensive option. By doing this, the higher cost option appears to be the better buy. Buyers will feel that they are getting the better deal and purchase the more expensive option, which is what you wanted in the first place. So, when presented with two options buyers tend to buy the cheaper option, but when a third strategically option is placed closer to the higher price option, the buyer will purchase the more expensive option.
- Online pricing offers an excellent example of the decoy effect. Here is a case from a leading magazine. Here is the setup from the magazine with three options for their potential subscribers. Option 1. a digital subscription ($50), Option 2. a print subscription ($125), Option 3. Both a print + digital subscription ($125). They are

focusing you on the third option they wanted to in the beginning.

- You can use an optical illusion with the decoy effect; it involves where you place of the decoy. Because most buyers will pick the middle option you have to set the decoy closer to the more expensive option it appears to be just a small increase in money for a much bigger return.
- Now could you use the middle position equally separated from one another options? The answer is by placing the more expensive and preferred option in the middle position.

Related to: Asymmetric Dominance effect

Framing: Using an approach or description of the situation or issue that is too narrow. Also, drawing different conclusions based on how data is presented. You can state an argument, marketing, discussion question in a positive way, or negative way, or a neutral way. The framing will affect the outcome or the answer.

Category: Belief and Decision Bias

Example:

- Positive vs. negative framing only 3 people passes vs. at least 258 failed!
- Participants saw a film of a traffic accident and then answered questions about the event, including the question 'About how fast were the cars going when they contacted each other?' Other participants received the same information, except that the verb 'contacted' was replaced by either *hit, bumped, collided*, or *smashed*. Even though all of the participants saw the same film, the wording of the questions affected their answers. The speed estimates (in miles per hour) were 31, 34, 38, 39, and 41, respectively.
  - One week later, the participants were asked whether they had seen broken glass at the accident site. Although the correct answer was 'no,' 32% of the participants who were given the 'smashed' condition said that they had.

Hence the wording of the question can influence their memory of the incident.

- A noted keystone of sales and social engineering.
  - o Changing the frame changes the context, which changes our interpretation, and consequently our experience.
  - o The more knowable and trained a person is in a specified field the more likely they are to make irrational, including life and death, decisions than an individual who is new or less indoctrinated to the specified field. That individual who are ingrained in their field will base their decisions on the superficial wording of the information rather than the real facts and probabilities. Someone who has been trained in cross-cultural and cross-discipline fields will analyze the facts not the word-tree of the speaker.

Fundamental Attribution: The tendency for people to over-emphasize personality-based explanations for behaviors observed while under-emphasizing the role and power of situational influences on the same behavior. Walk a mile in the other person's shoes. Which means to truly understand a person's reasoning, behavior, or reaction you must see the situation from their point of view. When we judge others' actions, we tend to give too much weight to their character and not enough to the circumstances in which they acted. While no accepted explanation is shared by everyone for this cognitive bias numerous hypotheses have been offered up for the cause of this error in human thinking. Interesting, this is one of the few cognitive thinking error that does have cultural differences that affect the explanation of this error. The research conducted has shown that groups from an "individualistic" culture are more inclined to exhibit the Fundamental Attribution, which groups from a collectivistic culture are not as likely to show this error in thinking.

Category: Social/Attributional Bias

Example:
- A classic example is a person who goes for their driving test. They fail and blame the situation, the car, the tester, the road conditions, etc. but not themselves, their lack of studying, their lack of preparation. Now, if they had

passed the exam, then if would have been that they are great and wonderful.
- If Alice saw Bob trip over a rock and fall, Alice might consider Bob to be clumsy or careless (personal/dispositional). But if Alice tripped over the same rock herself, she would be more likely to blame the placement of the rock (situational).

See also: Actor-Observer Bias, Group Attribution Error or Attribution Effect, Positivity Effect, Negativity Effect, or Correspondence bias.

Gambler's Fallacy: The tendency to think that future probabilities are altered by past events when in reality they remain unchanged. The belief that events play out on a fixed ratio of occurrence. The Journal of Risk and Uncertainty defined, in 1994, the gambler's fallacy as "the belief that the probability of an event is decreased when the event has occurred recently."

Category: Belief and Decision Bias

Example:
- Flipping a coin is always 50/50 that it will be heads or tails but believing that one surface will be greater than the other based on past performance.
- Lady Luck is smiling on me tonight.
- In playing the roulette table the last four spins of the wheel had the ball land on the color black, you may want to believe that the next turn will land the ball on a red color. Just the opposite, you may think the ball will continue to fall on the color black.
- During the middle age, people would gamble that the birth of a child would be a boy or girl. They had no idea about how the sex of a child was determined.

See also: Monte Carlo fallacy, the negative recency effect, or the fallacy of the maturity of chances

Hyperbolic Discounting: is the tendency for people to have a stronger preference for more immediate payoffs relative to later payoffs. Hyperbolic discounting leads to choices that are inconsistent over time – people make choices today that their future selves would prefer not to have made, despite using the same reasoning. This is hyperbolic discounting in action.

Category: Belief and Decision Bias

Example:
- NOW is always better than later.
- There was an important research in experimental psychology study (Stanford) in delayed gratification where the researcher "Offer a toddler a piece of candy now or two pieces of candy 15 minutes later.
- It was shown in adults that hyperbolic discounting doesn't change much as we mature.

Related to: Current Moment Bias, Present-Bias, Dynamic Inconsistency.

IKEA effect: The tendency for people to place a disproportionately high value on objects that they partially assembled themselves, such as furniture from IKEA, regardless of the quality of the result.

Category: Belief and Decision Bias

Example:
- Involving everyone in the company in making decision will share the ownership of the project.
- Every time you see the pride of the worker, who say something like, " I contributed to making that," you see the IKEA effect.

Illusory Truth Effect: A self-reinforcing process in which a collective belief gains more and more plausibility through its increasing repetition in public discussions or reports.

Category: Belief and Decision Bias

Example:
- Keep repeating something long enough, and it will become real.
- Political slogans and campaign rhetoric.
- Catchy advertising slogans and jingles are used to make us pay attention and repeat the advertiser's message to ourselves. (See Rhyme-As-Reason.)

Related to: Also known as the **truth effect** or the **illusion-of-truth effect**

Loss Aversion Bias: Loss aversion is the human tendency to try to avoid loss over acquiring gain. In colloquial term, it is worse to lose one's jacket than to find one. It has been suggested that psychologically, losses are twice as powerful. Once burned, twice shy. Neuroscience points out that humans may be hardwired to be loss averse due to our psychobiological evolution and its pressure on gains and losses.

Category: Belief and Decision Bias

Example:
- In the early days of humanoids, the loss of a day's food could amount to death. It appears that "serial gamblers" are the exception to loss aversion.

See also: Sunk cost, Endowment Effect

Ostrich effect: Ignoring an obvious negative situation.

Category: Belief and Decision Bias

Example:
- That check engine light that is glowing can't be ignored for long.
- Those attacks on the network can't be overlooked.
- Insider threats must be taken seriously.

Related to: Loss Aversion, Elephant in the Room

Overconfidence Bias: Excessive confidence in one's own answers to questions or ability to solve a problem. This is a well-established and much studied cognitive bias. Individuals will identify subjective confidence in their judgments is greater than the objective accuracy in reality. The best example is that people who rate their answer as 99% sure are incorrect 40% of the time.

In the literature overconfidence can be divided into three distinct ways as stated in Wikipedia:
1. Overestimation of one's actual performance;
2. Overplacement of one's performance relative to others;
3. Overprecision in expressing unwarranted certainty in the accuracy of one's beliefs

The Overconfidence Bias can be shown in many different major areas.
1. Overestimation
2. Over-precision
3. Over-placement

Overconfidence is called the most "pervasive and potentially catastrophic" of all the cognitive biases. It has been the source of countless lawsuits, strikes, wars, and stock market bubbles and crashes, to list just a few.

Category: Belief and Decision Bias

Example:

- In playing trivia, my teammates will ask are you sure? And I say yes, I'm 99% sure but I'm wrong 40% of the time.
- Many times you will hear people state "tell me the best choice or I don't have time to weed out the options. To help with this thinking have two options to explore before making a decision.
- Use a decision team. Multiple people are exploring the options from various angles and possibilities.
- Use a systematic approach to studying and making decisions, not you "gut feeling." Use this framework in making decisions and modify the structure to refine your and your's team approach.
- There is nothing wrong with bringing in "outsiders" to help analyze a decision or review your decision framework.

Related to: Dunn-Kruger effort

Rhyme-as-Reason: is a cognitive bias where a saying or an aphorism is judged as more accurate or truthful when it is rewritten to rhyme. If you want to leave a lasting impression of your representation or, of an important idea, develop a rhyme of the key point.

Category: Belief and Decision Bias

Example:
- "If the gloves don't fit, then you must acquit."
- "Loose lips sink ships."
- "Red sky at morning, sailors take warning."
- "Birds of the feather, flock together."
- "Leaves of three, let it be."

Related to: Eaton-Rosen phenomenon, Illusory Truth Effect

Prospect Theory: Human love the status quo, we want certainly in our life, day-after-day the same thing within a narrowly defined range. We will sacrifice income opportunities for certainty. For close to 400 hundred years classical economist believed and developed theories, algorithms based on the idea that human-made rational decisions. In the early to mid- seventy a group of cognitive scientist espoused the idea that human is irrational beings who are influenced by our environment and the interpretation of our environment much more than we would like to believe. Because of this, we make decisions that are not in our best interest. Behavioral economics breaks down the human decision-making process. Using Prospect theory, we begin to understand the factors that affect decision making and the choices we make.

Several ideas that Prospect Theory and Behavioral economics bring that helps in understand cognitive biases and how to apply cognitive biases to marketing, sales, and social engineering includes:
- "Incentives are the cornerstone of modern life."
- "Knowing what to measure, and how to measure it, can make a complicated world less so."
- "Conventional wisdom is often wrong."
- "Correlation does not equal causality."

From the work of Steven Levitt and Stephen Dubner.

Category: Not a bias, but a cornerstone of Behavioral Economics and related to several biases.

Example:
- Staying with the status quo when it would be best to go with something new
- Taking out loans that we cannot afford
- Smoking though we know it is harmful
- Purchasing items things we do not need
- Saving too little for a "raining day" or retirement.

Related to: Ambiguity, Loss Aversion, and several others.

Peak-End Rule: We are willing to sacrifice our comfort as long as there is a positive outcome in sight. We will remember the apex and the end. Memories of the whole will be determined by the feeling you had at the moment your brain considered to be the crest and the moment of the ending. It's not that your brain forgets the information, it that the brain filter out and conscience on the peak and the end of an event.

Category: Belief and Decision Bias

Example:
- Taking out loans that we cannot afford
- Music is a great example of the Peak-End Rule. While writing this book, we had Beethoven's Saints' Day Overture playing. It's easy to see and feel the climactic point that invites the listener to become involve the music, then it ends on a beautiful positive chord, making one want to say, out loud, where can I find more of that music?
- Another example is the old-time preacher who knew how to build the sermon to a fever pitch that everyone could remember, then softly and positively bring the closing with the inspirational message that you could be one of the chosen, ending with a great invitational song, with words like "come home, come home...."

Projection Bias: is a feature in human thinking where one thinks that others have the same priority, attitude or belief that one harbors oneself, even if this is unlikely to be the case. In the article written by Dvorsky, he discusses that "we tend to assume that most people think just like us - though there may be no justification for it." We have a tendency to overestimate how typical and normal we are, and assume that a consensus exists on matters when there is no real consensus. This bias appears in the group that assumes more people on the outside of the group agree or are like them than is the case.

Category: Social/Attributional Bias

Example:
- I knew a brother and sister who were from the US but grew up in India. All the fathers were engineers of some type. All the mothers were also college graduates. The group living in this compound were from many different countries. From this small group, the image is that everyone from the US goes to college. I show this brother and sister the real statistics for college-bound individuals in the US; they were shocked how reality was different from the projection they grew up with in India.

Status Quo Bias: The tendency for people to like things to stay relatively the same. Know as the "Steady State" in economics.

Category: Belief and Decision Bias

Example:
- Comfort zone think of HGTV International House Hunters. Watch the TV show where people move oversea to another country, and they want the new house or apartment to be just like the one they left in their countries of origin.
- Want an individual to use your brand of browser or email, then have it automatically install on the system and be the default for that system. Most people will just use what's is installed. People will defend the default option, not at would already there, but it the best, because it was the default.
- Always give people a default option plan when offering choices. Individuals are naturally predisposed to select the default. Frame your presentation with a default option.

Related to: loss aversion, endowment effect, and system justification biases

7 METHODS USED TO CHALLENGE COGNITIVE BIASES

Developing methodologies to challenge cognitive biases is a lot like being an "armchair quarterback." We have to start with past projects, laboratory failures, real-world effects of cognitive biases. Roberta Wohlstetter found it easy after the events of Pearl Harbor to sort the relevant from the irrelevant signals. "After the event, of course, a signal is always crystal clear; we can now see what disaster it was signaling since the catastrophe has occurred. But before the event, it is obscure and pregnant with conflicting meaning." (Jones)

Scientific Method: The authors believe that the best approach or system for vetting and limiting the consequences of cognitive bias is the scientific method, as it places evidence and methodology behind an idea under open scrutiny. By this, many opinions and separate analyses can be used to compensate for the bias of any one individual. It is important to remember, however, that in everyday life just knowing about these biases doesn't necessarily free you from them. (Hanson)

Key Assumptions Check: Is most useful at the beginning of an analytic project. (CIA) Later on, during this project, you can revisit the Key Assumptions Check.

An example of how using the Key Assumptions Check during the 2002 DC sniper case could have allowed law enforcement officials to:

- Avoid jumping to the following conclusions by explicitly examining each assumption:
  - ☐ the shooter is white
  - ☐ has military training
  - ☐ is driving a white van
- Be receptive to new leads and tips
- More seriously considered evidence that subsequently became available, that contradicted a fundamental assumption.
  - ☐ Many times key assumptions get lost in the noise, leading investigator to not think about challenges to this assumption.

Continuing from the Tradecraft Primer, "Checking for the major assumptions requires analysts to consider how their analysis depends on the validity of certain premises which they do not routinely question or believe to be in doubt."

How to challenge assumptions was the subject of a study by (Roberto) that "explicitly focused on the cognitive bias as a potential contributor to a disaster-level event; this study examined the causes of the loss of several members of two expedition teams on Mount Everest on two consecutive days in 1996. This study concluded that several cognitive biases were 'in play' on the mountain, along with other human dynamics. This was a case of highly trained, experienced people breaking their rules, apparently under the influence of the overconfidence effect, the sunk cost fallacy, and the availability heuristic. Five people, including both expedition leaders, lost their lives despite explicit warnings in briefings before and during the ascent of Everest. In addition to the leaders' mistakes, most team members, though they recognized their leader's faulty judgments, failed to insist on following through on the established ascent rules." I (Hale) would add that the followers were exhibiting the following cognitive biases: confirmation bias, bandwagon effect, groupthink, and herd behavior.

The four-step process for the Key Assumption Checklist includes the following:

1. Review what the current analytic line on this issue appears to be; write it down for all to see.
2. Articulate all the premises, both stated and unstated in finished intelligence, which is accepted as true for this analytic line to be valid.

3. Challenge each assumption, asking why it "must" be true and remains valid under all conditions.
4. Refine the list of key assumptions to contain only those that "must be true" to sustain your analytic line; consider under what conditions or in the face of what information these assumptions might not hold.

Following is a list of questions which should be asked during the development of the Key Assumption Checklist.

- How much confidence exists that this assumption is correct?
- What explains the degree of confidence in the assumption?
- What circumstances or information might undermine this assumption?
- Is a key assumption more likely a key uncertainty or key factor?
- Could the assumption have been true in the past but less so now?
- If the assumption proves to be wrong, would it significantly alter the analytic line?
- Has the Key Assumption Checklist process identified new factors that need further analysis?

In closing our discussion of the Key Assumption Checklist, The authors would direct your attention to Pherson Associates ACH Software Tool (www.pherson.org).

In their presentation for HSDECA entitled "Use of
Analytic Tools and Techniques in the Homeland
Security Classroom," Pherson Associates
emphasized the following:

Five classic analytic traps

1. If we don't have a category for something,
   we usually ignore it.
2. We discount facts that do not support our
   analysis.
3. We overstate conclusions when a little data
   is consistent.
4. We do not change our analysis despite
   mounting contradictions.
5. We assume the present is like the past.

Key Analytic Techniques

- Challenge your assumptions
- Generate multiple hypotheses
- Search for inconsistent data
- Check the reliability of the key evidence
- Develop indicators

Cognition Method: Campbell states in her article
"Battling Cognitive Bias" that "all is not lost – there
are ways to lessen the effect of cognitive biases in
our work." She goes on to list and discuss possible
remedies which I will restate below:

1. **Awareness:** If you are not conscious of cognitive biases, then you will not be able to challenges then. Campbell recommends the approach of looking at examples of cognitive biases from famous cases. Sent you are not personally involved in these cases you can be more critical in examining these cases.
2. **Review the Past:** "This may be the hardest step. This involves reviewing your previous work and decisions and trying to seek out these tendencies. Not everyone is as equally affected, and finding out which biases are Battling Cognitive Bias your particular weaknesses can help you focus your efforts in the future." (Campbell)
3. **Change the People:** "Sometimes, the only way to have an impartial review of a project or decision is to bring in other people who were not involved in the original decisions. The sunk cost effect is easy to avoid," Yes, sunk cost is another cognitive bias, "if one wasn't the person who sunk those costs, to begin with. Bringing in outside consultants can be useful, but again one needs to be careful that those hiring the consultants and deciding on their pay (and whether said consultants will be hired for future projects) aren't going to influence the consultants to come up with a foregone conclusion." (Campbell)
4. **Change the Group Dynamics or Composition:** "As opposed to taking outside people to replace the ones who made the decision, sometimes it's enough to mix up insiders with diverse areas of expertise and experience."

In researching scientific problem-solving, Kevin Dunbar noted a difference in two labs, both of which had the same experimental problem that needed solving. One of the two labs solved their problem much more quickly — the lab that had a more diverse composition regarding expertise. The faster-solving group had biochemists, geneticists, graduate students, and molecular biologists; they all had different training and different perspectives going into the problem-solving process. The slower-solving group was composed solely of E. coli experts sharing the same assumption sets and the same training.

"It's hard to suffer confirmation bias when group members have different positions they're trying to confirm. Individually, people may have problems, but as long as they're not all aligned in the same direction, the diversity of thought can help solve problems better." (Campbell)

Check Assumptions and Actively Seek Disconfirming Evidence "Perhaps you are not solving in the group, but alone. Your perspective is all you've got, and so some of the above fixes may not be open to you. However, if you turn your implicit assumptions into an explicit list, and actively try to see if your assumptions are wrong, you can combat confirmation bias. It does require progressive discipline and a willingness to find your assumptions flawed." The above recommendations come from the research developed by Lehrer and Roberto, see references at the end of the book for links, as reported by Campbell.

Summary:

Even though the authors have their preferences, we recommend a combination approach in challenging cognitive biases. Practice using different techniques come up with your combination, and then throw in another every now and again to double check yourself.

8 PARTIAL LISTING OF CHARTS USED DURING OUR PRESENTATIONS

The authors have presented the contents of this book to diverse audiences. Generally, we showed how cognitive biases both challenge and affect security and cyber warfare. The following is a partial display of those charts giving hopefully a taste of our presentation to the reader.

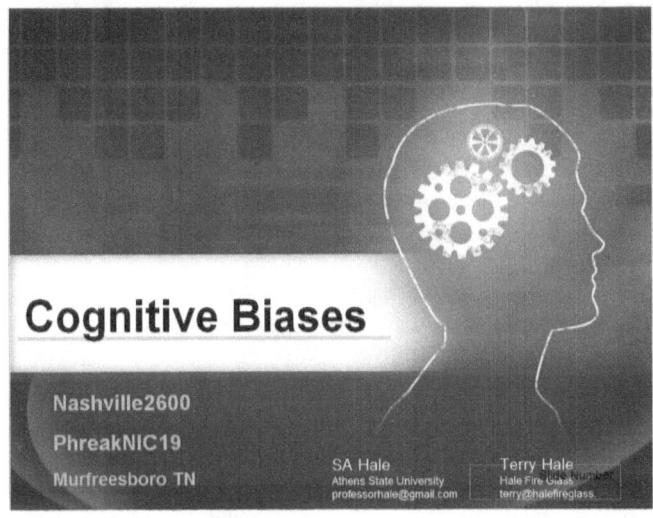

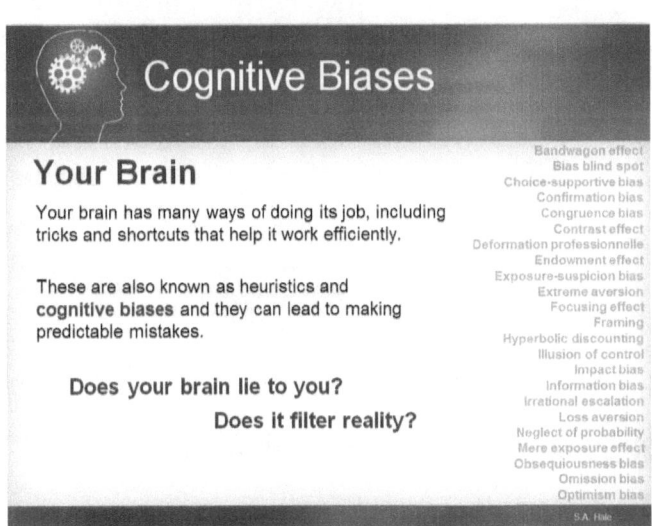

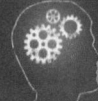

Cognitive Biases

definitions

...the inherent thinking errors that Homo Sapiens make in processing information.

...thinking errors that prevent one from accurately understanding reality, despite all the needed data and evidence.

...help us with simple, concrete day-to-day problems -- not highly mathematical or abstract problems.

S.A. Hale

Cognitive Biases

Theoretical Cause

System 1, System 2 Model

System 1 consists of heuristic shortcuts developed over evolutionary time and implemented in certain brain cortices as a kind of Heuristic Toolkit.

System 2 is the evolutionarily modern component of our reasoning mechanism, which operates serially, slowly, and (partly) visibly to our conscious minds.

S.A. Hale

 Cognitive Biases

Causation

Despite several decades of research and experimentation, no comprehensive theory of what creates or causes cognitive biases has emerged.

It's just part of being "human."

S.A. Hale

 Cognitive Biases

Neuroscience Research

supports the idea that the interaction between

System 1 and System 2

is what triggers the

Cognitive Bias

S.A. Hale

Cognitive Biases

Groupings

Decision Making — The deviation from what is normatively expected.

Social (Attribution) — People involved in an action (*actors*) view things differently from people not involved (*observers*).

Memory — Either enhances or impairs the recall of a memory or alters the content of a reported memory.

S.A. Hale

Cognitive Biases

Focused Cognitive Biases for BICDW

Decision Making Biases

Anchoring	• Relying too heavily on a past reference or on one trait or piece of information when making decisions.
Availability heuristic	• Estimating by memory, biased toward vivid, unusual, or emotionally charged examples.
Bandwagon effect	• Doing/believing things because many other people do/believe the same. *Group think* or *herd behavior*.
Bias Blind Spot	• Seeing oneself as less biased than other people, or identifying more biases in others than in oneself.
Confirmation bias	• Searching for or interpreting information in a way that confirms one's preconceptions.
Overconfidence effect	• Excessive confidence in one's own answers to questions. "99% certain" is wrong 40% of the time.

S.A. Hale

Cognitive Biases
Focused Cognitive Biases for BICDW

Social (Attribution) Biases

Actor-Observer	• Actions of others are based on personality while our own actions are based more on situation.
Defensive Attribution	• In a mishap, defensiveness of those responsible will depend upon the severity of the outcomes .
Dunning-Kruger Effect	• Failing to realize one's own incompetency due to inability to distinguish between competence and incompetence
Out-group Homogeneity	• Seeing members of their own group as being relatively more varied than members of other groups.
System Justification	• The tendency to defend and bolster the status quo.
Ultimate Attribution Error	• A person is likely to make an internal attribution to an entire group instead of the individuals within the group.

S.A. Hale

Cognitive Biases
Focused Cognitive Biases for BICDW

Memory Biases

Von Restorff effect	• Isolation effect; remembering the unusual.
Telescoping Effect	• Displacing recent events backward in time and remote events forward in time.
Self-Serving Bias	• Failing to realize one's own incompetency due to inability to distinguish between competence and incompetence
Google Effect	• The tendency to forget information that can be easily found online.
Misattribution Bias	• Information is retained in memory but the source of the memory is forgotten.
Egocentric Bias	• Recalling the past in a self-serving manner.

S.A. Hale

Cognitive Biases

Hindsight Bias

"After the event, of course, a signal is always crystal clear; we can now see what disaster it was signaling since the disaster has occurred. But before the event it is obscure and pregnant with conflicting meaning."

~ Lloyd Jones

Pattern of Error: Perceptual and Cognitive Bias in Intelligence Analysis and Decision-Making

Cognitive Biases

Challenging Cognitive Biases

Scientific Method

- Probably the best way for vetting and limiting the consequences of cognitive bias.

- Places evidence and methodology behind idea under open scrutiny.

- Many opinions and separate analyses can be used to compensate for the bias of any one individual.

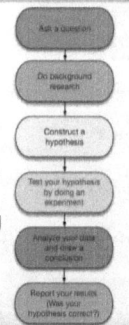

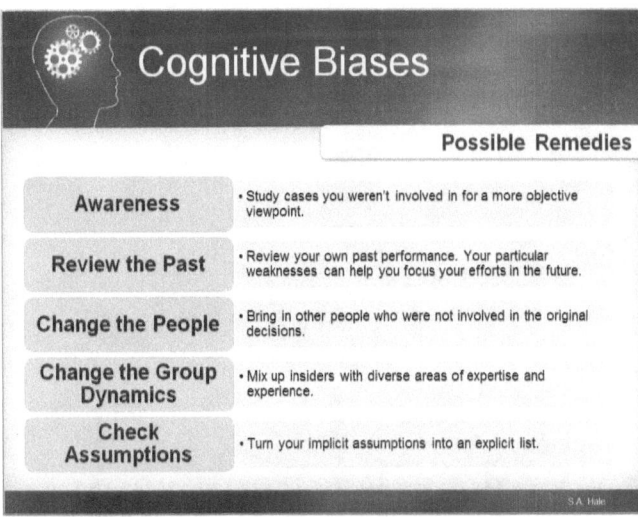

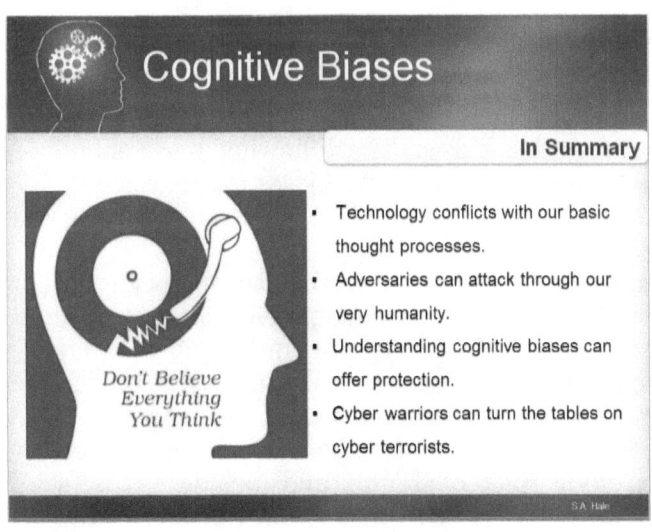

9 COGNITIVE BIASES GUIDE FROM THE ROYAL SOCIETY OF ACCOUNT PLANNING

In this book we, the authors, have shared cognitive biases that we have had experiences with in our work. What follows is a detailed listing of hundred or so cognitive biases. The following is a general discussion of those cognitive biases for your benefit and exploration.

A group of researchers at the Royal Society of Account Planning defined cognitive biases, In the Visual Study Guide to Cognitive Biases, as "psychological tendencies that cause the human brain to draw incorrect conclusions. Such biases are thought to be a form of "cognitive shortcut," often based upon rules of thumb, and include errors in statistical judgment, social attribution, and memory. These biases are a common outcome of human thought, and often drastically skew the reliability of anecdotal and legal evidence. The phenomenon is studied in cognitive science and social psychology."

The following is a list of the cognitive biases with a short definition and examples of different cognitive biases.

9.1 Social Biases

Definition and examples of Social Biases

Forer effect / Barnum effect
The tendency to give high accuracy ratings to descriptions of their personality that supposedly are tailored specifically for them, but are in fact vague and general enough to apply to a wide range of people. For example, horoscopes.

Ingroup bias
The tendency for people to give preferential treatment to others they perceive to be members of their groups.

Self-fulfilling prophecy
The tendency to engage in behaviors that elicit results which will (consciously or not) confirm existing attitudes.

Halo effect
The tendency for a person's positive or negative traits to "spill over" from one area of their personality to another in others' perceptions of them (see also physical attractiveness stereotype).

Ultimate attribution error
Similar to the fundamental attribution error, in this error, a person is likely to make an internal attribution to an entire group instead of the individuals within the group.

False consensus effect
The tendency for people to overestimate the degree to which others agree with them.

Self-serving bias or Behavioral confirmation effect
The tendency to claim more responsibility for successes than failures. It may also manifest itself as a tendency for people to evaluate ambiguous information in a way beneficial to their interests (see also group-serving bias).

Notational bias
A form of cultural bias in which the notational conventions of recording data biases the appearance of that data toward (or away from) the system upon which the notational schema is based.

Egocentric bias
Occurs when people claim more responsibility for themselves for the results of a joint action than an outside observer would.

Just-world phenomenon
The tendency for people to believe that the world is just and therefore people "get what they deserve."

System justification effect / Status Quo Bias

The tendency to defend and bolster the status quo. Existing social, economic, and political arrangements tend to be preferred, and
alternatives disparaged sometimes even at the expense of individual and collective self-interest. (See also status quo bias.)

Dunning-Kruger / Superiority Bias
Overestimating one's desirable qualities, and underestimating undesirable qualities, relative to other people. Also known as Superiority bias
(also known as "Lake Wobegon effect," "better-than-average effect," "superiority bias," or Dunning-Kruger effect).

Illusion of asymmetric insight
People perceive their knowledge of their peers to surpass their peers' knowledge of them.

Herd instinct
Common tendency to adopt the opinions and follow the behaviors of the majority to feel safer and to avoid conflict.

Illusion of transparency
People overestimate others' ability to know them, and they also overestimate their ability to know others.

Fundamental attribution error / Actor-observer bias
The tendency for people to over-emphasize personality-based explanations for behaviors observed in others while underemphasizing the role and power of situational influences on the same behavior (see also actor-observer bias, group

attribution error, positivity effect, and negativity effect).

Projection bias
The tendency to unconsciously assume that others share the same or similar thoughts, beliefs, values, or positions.

Outgroup homogeneity bias
Individuals see members of their group as being relatively more varied than members of other groups.

Trait ascription bias
The tendency for people to view themselves as relatively variable in terms of personality, behavior, and mood while viewing others as much more predictable.

9.2 Memory Biases

Definition and examples of Memory Biases

Suggestibility
A form of misattribution where ideas suggested by a questioner are mistaken for memory.

Reminiscence bump
The effect that people tend to recall more personal events from adolescence and early adulthood than from other lifetime periods.

Cryptomnesia / False memory

A form of misattribution where memory is mistaken for imagination, or the confusion of true memories with false memories.

Consistency bias
Incorrectly remembering one's past attitudes and behavior as resembling present attitudes and behavior.

Rosy retrospection
The tendency to rate past events more positively than they had rated them when the event occurred.

Self-serving bias
Perceiving oneself responsible for desirable outcomes but not responsible for undesirable outcomes.

Egocentric bias
Recalling the past in a self-serving manner, e.g. remembering one's exam grades as being better than they were, or remembering a caught fish as being bigger than it was.

Hindsight bias
Filtering memory of past events through present knowledge, so that those events look more predictable than they were; also known as the 'I-knew-it-all-along effect.' "The grapes are sour anyway."

9.3 Decision-making Biases

Definition and examples of Decision-making Biases

Hyperbolic discounting
The tendency for people to have a stronger preference for more immediate payoffs relative to later payoffs, where the tendency increases the closer to the present both payoffs are.

Irrational escalation
The tendency to make irrational decisions based upon rational decisions in the past or to justify actions already were taken.

Mere exposure effect
The tendency for people to express undue liking for things merely because they are familiar with them.

Omission bias
The tendency to judge harmful actions as worse, or less moral, than equally harmful omissions (inactions).

Mere exposure effect
The tendency for people to express undue liking for things merely because they are familiar with them.

Negativity bias
The phenomenon by which humans pay more attention to and give more weight to negative than positive experiences or other kinds of information.

Neglect of probability
The tendency to completely disregard probability when making a decision under uncertainty.

Normalcy bias
The refusal to plan for, or react to, a disaster which has never happened before.

Interloper effect / Consultation paradox
The tendency to value third party consultation as objective, confirming, and without motive. Also consultation paradox, the conclusion that solutions proposed by existing personnel within an organization are less likely to receive support than from those recruited for that purpose.

Focusing effect
Prediction bias occurring when people place too much importance on one aspect of an event; causes error in accurately predicting the utility of a future outcome.

Illusion of control
The tendency for human beings to believe they can control or at least influence outcomes that they clearly cannot.

Outcome bias
The tendency to judge a decision by its eventual outcome instead of based on the quality of the decision at the time it was made.

Post-purchase rationalization
The tendency to persuade oneself through rational argument that a purchase was a good value.

Framing
Using an approach or description of the situation or issue that is too narrow. Also framing effect – Drawing different conclusions based on how data is presented.

Experimenter's or Expectation bias

The tendency for experimenters to believe, certify, and publish data that agree with their expectations for the outcome of an experiment, and to disbelieve, discard, or downgrade the corresponding weightings for data that appear to conflict with those expectations.

Information bias
The tendency to seek information even when it cannot affect action.

Extraordinary bias
The tendency to value an object more than others in the same category as a result of an extraordinary of that object that does not, in itself, change the value.

Planning fallacy
The tendency to underestimate task-completion times.

Semmelweis reflex
The tendency to reject new evidence that contradicts a traditional paradigm.

Déformation Professionnelle
The tendency to look at things according to the conventions of one's profession, forgetting any broader point of view.

Not Invented Here
The tendency to ignore that a product or solution already exists, because its source is seen as an "enemy" or as "inferior."

Impact bias

The tendency for people to overestimate the length or the intensity of the impact of future feeling states.

Moral credential effect
The trend of a track record of non-prejudice to increase subsequent prejudice.

Base rate fallacy
Ignoring available statistical data in favor of particulars.

Bias blind spot
The tendency not to compensate for one's cognitive biases

Confirmation bias
The tendency to search for or interpret information in a way that confirms one's preconceptions.

Congruence bias
The tendency to test hypotheses exclusively through direct testing, in contrast to tests of possible alternative hypotheses.

Contrast effect
The enhancement or diminishing of weight or another measurement when compared with a recently observed the contrasting object.

Denomination effect
The tendency to spend more money when it is denominated in small amounts (e.g. coins) rather than large amounts (e.g. bills).

Distinction bias

The tendency to view two options as more dissimilar when evaluating them simultaneously than when evaluating them separately.

Endowment effect / Loss aversion
"the fact that people often demand much more to give up an object than they would be willing to pay to acquire it." (see also sunk cost effects)

Choice-supportive bias
The tendency to remember one's choices as better then they were

Bandwagon effect
The tendency to do (or believe) things because many other people do (or believe) the same. Related to groupthink and herd behavior.

Selective perception
The tendency for expectations to affect perception.

Reactance
The urge to do the opposite of what someone wants you to do out of a need to resist a perceived attempt to constrain your freedom of choice.

Money illusion
The tendency of people to concentrate on the nominal (face value) of money rather than the value regarding purchasing power.

Zero-risk bias

Preference for reducing a small risk to zero over a greater reduction in a larger risk.

Wishful thinking
The formation of beliefs and the making of decisions according to what is pleasing to imagine instead of by appeal to evidence or rationality.

Need for Closure
The need to reach a verdict in important matters; to have an answer and to escape the feeling of doubt and uncertainty. The personal context (time or social pressure) might increase this bias.

Restraint bias
The tendency to overestimate one's ability to show restraint in the face of temptation.

Von Restorff effect
The tendency for an item that "stands out like a sore thumb" to be more likely to be remembered than other items

Pseudocertainty effect
The tendency to make risk-averse choices if the expected outcome is positive, but make risk-seeking choices to avoid negative outcomes.

Status quo bias
The tendency for people to like things to stay relatively the same (see also loss aversion, endowment effect, and system justification).

9.4 Probability or Belief Biases

Definition and examples of Probability or Belief Biases

Positive outcome bias
The tendency to overestimate the probability of good things happening to them (see also wishful thinking, optimism bias, and valence effect).

Disregard of regression toward the mean
The tendency to expect extreme performance to continue.

Selection bias
A distortion of evidence or data that arises from the way that the data are collected.

Survivorship bias
The tendency to concentrate on the people or things that "survived" some process and ignoring those that didn't, or arguing that a strategy is effective given the winners, while ignoring a lot of losers.

Telescoping effect
The effect that recent events appear to have occurred more remotely and remote events appear to have taken place more recently.

Texas sharpshooter fallacy
The fallacy of selecting or adjusting a hypothesis after the data is collected, making it impossible to test the hypothesis fairly. Refers to the concept of firing shots at a barn door, drawing a circle around the best group, and declaring that to be the target.

Pareidolia

A vague and random stimulus (often an image or sound) is perceived as significant, e.g., seeing images of animals or faces in clouds, the man in the moon, and hearing hidden messages on records played in reverse

Outcome bias
The tendency to judge a decision by its eventual outcome instead of based on the quality of the decision at the time it was made.

Overconfidence effect
Excessive confidence in one's answers to questions. For example, for certain types of question, answers that people rate as "99% certain" turn out to be wrong 40% of the time.

Observer-expectancy effect
When a researcher expects a given result and therefore unconsciously manipulates an experiment or misinterprets data in order to find it (see also subject-expectancy effect).

Hindsight bias
Sometimes called the "I-knew-it-all-along" effect, the tendency to see past events as being predictable.

Hawthorne effect
The tendency to perform or perceive differently when one knows they are being observed.

Gambler's fallacy

The tendency to think that future probabilities are altered by past events when in reality they are unchanged. Results from an erroneous conceptualization of the law of large numbers. For example, "I've flipped heads with this coin five times consecutively, so the chance of tails coming out on the sixth flip is much greater than heads."

Clustering illusion
The tendency to see patterns where none exist.

Illusory correlation
Beliefs that inaccurately suppose a relationship between a particular type of action and an effect.

Last illusion
The belief that someone must know what is going on.

Attentional bias
The tendency to neglect relevant data when making judgments of a correlation or association.

Availability heuristic
Estimating what is more likely by what is more available in memory, which is biased toward vivid, unusual, or emotionally charged examples.

Belief bias
An effect where someone's evaluation of the logical strength of an argument is biased by the believability of the conclusion.

Ostrich effect
Ignoring an obvious (negative) situation.

Ambiguity effect
The tendency to avoid options for which missing information makes the probability seem "unknown."

Availability cascade
A self-reinforcing process in which a collective belief gains more and more plausibility through its increasing repetition in public discourse (or "repeat something long enough, and it will become true").

Capability bias
The tendency to believe that the closer average performance is to a target, the tighter the distribution of the data set.

Conjunction fallacy
The tendency to assume that specific conditions are more probable than general ones.

Authority bias
The tendency to value an ambiguous stimulus (e.g., an art performance) according to the opinion of someone who is seen as an authority on the topic.

Disposition effect
The tendency to sell assets that have increased in value but hold assets that have decreased in value.

Stereotyping
Expecting a member of a group to have certain characteristics without having actual information

about that individual.

Subadditivity effect
The tendency to judge the probability of the whole to be less than the probabilities of the parts.

Subjective validation
perception that something is true if a subject's belief demands it to be true. Also, assigns perceived connections between coincidences.

Well traveled road effect
Underestimation of the duration taken to traverse oft-traveled routes and over-estimate the duration taken to traverse less familiar routes.

Neglect of prior base rates effect
The tendency to ignore known odds when reevaluating odds in light of weak evidence.

Optimism bias
The tendency to be overoptimistic about the outcome of planned actions.

Primacy effect
The tendency to weigh initial events more than subsequent events.

Anchoring effect
The tendency to rely too heavily on, or "anchor on" a past reference or one trait or piece of information when making decisions (also called "insufficient adjustment").

Recency effect / Peak-end rule
The tendency to weigh recent events more than earlier events.

Cognitive Biases Affect on Security

9 SUMMARY

Don't believe everything you think!

With the age of technology conflicting with our basic thought processes, sales, marketing our fundamental thought processes. Adversaries can attack our cyber systems not only through routers, computers, and code but through our very humanity.

An understanding of the numerous cognitive biases, along with the techniques and methods discussed in this book, will help protect both our human and cyber defense systems by repelling unwanted advances and offering protection for our vulnerabilities.

Also, because cognitive bias is a cross-cultural, rudimentary part of being human, we can turn the tables on the unwanted sale people, marketers, and cyber terrorist.

10 REFERENCES

Aamodt, S. A. & Wang, S., (2009). *Welcome to Your Brain: Why You Lose Your Car Keys but Never Forget How to Drive and Other Puzzles of Everyday Life*. 1st ed. USA: Bloomsbury, USA, New York.

Anderson, Mike (unknown). <u>Hacking the Mind by Understanding Biases - Ambiguity Effect</u>. http://www.mikeyanderson.com

Ariely, D., (2008). *Predictably Irrational: The Hidden Forces that Shape Our Decisions*. 1st ed. New York, NY: HarperCollins.

Baron, J., (2007). *Thinking and deciding*. 4th ed. New York, NY: Cambridge University Press.

Benson, J., (2002). *Why Plans Fail: Cognitive Biases of Decision Making in Business*. 1st ed. USA: Modus Cooperandi Press.

Campbell, Mary, Pat, (2010). Battling Cognitive Bias. The Stepping Stones. Society of Actuaries.

CIA, (2009). *A Tradecraft Primer*. USA: Center for Study of Intelligence Progressive Management.

Conroy, P. (2012). *Cognitive Bias Mitigation*. [ONLINE] Available at http://pfconroy.wordpress.com/. [Last Accessed 6 August 2012].

Dougherty, M. R. P., Gettys, C. F., & Ogden, E. E., (1999). MINERVA-DM: A memory processes model for judgments of likelihood. *Psychological Review*. 106 (1), pp.180-209

Elster, Jon (1983). *Sour Grapes: Studies in the Subversion of Rationality*. Cambridge, UK: Cambridge University Press.

Festinger, L. (1957). *A theory of cognitive dissonance*. Stanford, CA: Stanford University Press.

Gigerenzer, Gerd; Selten, Reinhard (2002). *Bounded Rationality*. Cambridge: MIT Press.

Hanson, Robin (2008). In Bias, Meta is Max. [ONLINE] Available at http://www.overcomingbias.com. [Last Accessed 6 August 2012].

Hasleton, M. G. & Nettle, D., (2006). The Paranoid Optimist: An Integrative Evolutionary Model of Cognitive Biases. *Personality and Social Psychology Review*. 10 (1), pp.47-66.

Hilbert, M., (2012). Toward a synthesis of cognitive biases: How noisy information processing can bias human decision making. *Psychological Bulletin*. 138 (2), pp.211-237.

Jones, L., (2005). *Pattern of Error: Perceptual and Cognitive Bias in Intelligence Analysis and Decision-making*. 1st ed. Monterey, California: Navy Post Graduate School.

Kahneman, D. & Tversky, A., (1972). Subjective probability: A judgment of representativeness. *Cognitive Psychology*. 3, pp.430–454.

Lehrer, Jonah. (2009) "Accept Defeat: The Neuroscience of Screwing Up," Wired Magazine, Link: http://www.wired.com/magazine/2009/12/fail_ac cept_defeat/all/1.

Matlin, Margaret (2009). *Cognition*. Hoboken, NJ: John Wiley & Sons, Inc. pp.413.

Newell, Benjamin R.; David A. Lagnado, David R. Shanks (2007). *Straight choices: the psychology of decision making*. Routledge. pp.71–74.

Pherson Associates (2010) "Use of Analytic Tools and Techniques in the Homeland Security Classroom" Presentation to HSDECA.

Roberto, Michael A. The Art of Critical Decision Making. Lectures from The Teaching Company. Link: http://www.teach12.com/ttcx/coursedes-clong2.aspx?cid=5932.

Roberto, M. A. (2002). "Lessons from Everest: The Interaction of Cognitive Bias, Psychological, Safety and System Complexity." California Management Review (2002) 45(1): 136–158.

Servitor, Videre, (2012). *Ghosts Inside Your Head*. 1st ed. USA: Amazon Digital Services.

Tversky, A; Kahneman (1973). "Availability: A heuristic for judging frequency and probability." *Cognitive Psychology* **5** (1): 207–233. doi:10.1016/0010-0285(73)90033-9.

ABOUT THE AUTHORS

Mr. Hale is a Computer Scientist with over 30 years of experiences. Predominantly utilizing an interdisciplinary approach that brings the human side to technology by combining Cognitive and Behavioral Science, with Computer and Security Engineering. Mr. Hale works in various areas including Security Engineering, Information Assurance, System Administration, Cyber Warfare, Cognitive System Engineering, and Software Engineering.

Mr. Hale uses the human-centric computing approach during his normal day as a Cyber Technical Lead, System Administrator, and Information System Security Officer (ISSO) for several classified laboratories.

He has provided engineering and analysis support to government and commercial groups both foreign and domestic.

Additionally, Mr. Hale is an Adjunct Professor of Computer Science where he teaches courses in System Security Management, Digital Forensics, Cyber Ethics, and Javascript and other courses.

You may contact Mr. Hale at ProfessorHale@gmail.com or AbbyNormalResearch@gmail.com

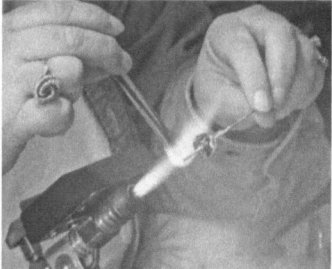

Terry Hale has been fascinated with the decision-making process since she began dealing with customers in her family's business when she was a young teen. Throughout her adult life, she has studied social engineering while working in a career that includes sales, teaching, corporate training, and NASA and Department of Defense technical support.

Currently, Terry is a full-time glass artist and jewelry designer. Her designs can be found in fine art galleries and shows throughout the southeast US. In addition to teaching in her studio, Terry regularly teaches at John C. Campbell Folk School in Brasstown NC, Appalachian Center for Craft (Tennessee Tech) in Smithville TN, and at Essence of Mulranny, County Mayo, Ireland. She depends on her study of cognitive biases while teaching and decision making while selling her artwork.

You contact Terry at terry@halefireglass.com or AbbyNormalResearch@gmail.com

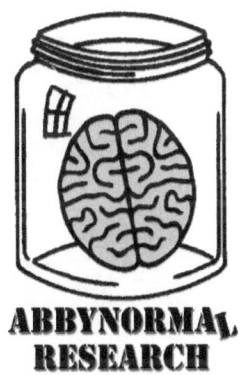

ABBYNORMAL
RESEARCH